屏住呼吸

Hold Your Breath

Gunter Pauli

[比]冈特·鲍利 著

[哥伦]凯瑟琳娜·巴赫 绘

章里西 译

上海遠東出版社

特别感谢以下热心人士对童书工作的支持：

匡志强　方　芳　宋小华　解　东　厉　云　李　婧

刘　丹　熊彩虹　罗淑怡　旷　婉　杨　荣　刘学振

何圣霖　王必斗　潘林平　熊志强　廖清州　谭燕宁

王　征　白　纯　张林霞　寿颖慧　罗　佳　傅　俊

胡海朋　白永喆　韦小宏　李　杰　欧　亮

目录

Contents

一头年轻的座头鲸正在温暖的海水中享受着日光浴。他妈妈看到他脸上沉迷的表情，问道：

“你恋爱了？”

“哦，妈妈，昨晚的舞蹈太棒了。”

A young humpback whale is enjoying basking in the warm waters of the ocean. His mother sees the dreamy expression on his face and asks:

"Have you fallen in love?"

"Oh Mom, that was such an amazing dance last night."

……舞蹈太棒了……

… such an amazing dance …

……月圆之夜见到你爸爸……

… met your dad under a full moon …

“啊，月光下的浪漫，多么可爱。我还记得那个月圆之夜见到你爸爸的情景……”

“妈妈，那只是一次跳舞好吧？但那又是多么美妙……”

“太棒了！那你约好再去见她了吗？”

“Ah, romance in the moonlight, how lovely. I do remember when I met your dad under a full moon…”

“It was only a dance, okay Mom? But what a wonderful one…”

“Great! So, did you make another date, to see her again?”

“没有，我的好舞伴很快就走了，我想她近期不会回来了。”

“这么说，你是和一个陌生人跳了舞——现在对你来说，她还是一个陌生人？”

“我希望不是。只跳了6分钟，但它太特别了。”

“No, my wonderful dancing partner left quickly, and I don't think she will be back anytime soon.”

“So, you danced with a stranger – who will now remain a stranger to you?”

“I do hope not. The dance only lasted six minutes, but it was so special.”

你是和一个陌生人跳了舞……

you danced with a stranger ...

我的舞伴必须屏住呼吸……

my dance partner has to hold her breath ...

“只跳了6分钟？我们那时候，要一直跳到凌晨。”

“我知道，但是我的舞伴必须屏住呼吸，所以能跳6分钟已经很了不起了。”

“嗯，我们鲸也必须屏住呼吸，但可以保持很长时间。对了，你昨晚和谁跳舞了？这个神秘女郎是谁？”

“Only six minutes? In my day, we danced until the early morning hours.”

“I know, but my dance partner has to hold her breath, so dancing for six minutes is already quite something.”

“Well, we whales also have to hold our breath, but we can do so for much longer. So, who were you dancing with last night? Who is this mysterious stranger?”

“如果我告诉你，这位漂亮的女孩是人类，你敢相信吗？”

“什么？儿子，你一定是梦见了不存在的美人鱼。陆地上的年轻女孩是不会和我们在水底跳舞的。她们不可能屏住呼吸那么久，尤其是在跳舞的时候！”

“大多数人都做不到，但是这个女孩很特别，妈妈。她解释说，当她们感到压力时，大脑会消耗掉她们血液中近一半的氧，所以她索性关闭了大脑……”

“Would you believe me if I told you this fine lady is a human?”

“What? You must have been dreaming, my boy, of mermaids that do not exist. Young ladies from the land do not come dancing under the water with us. They could never hold their breath for that long, especially not while dancing!”

“Most people would not be able to, but this one is special, Mom. She explained that since their brains consume almost half of all the oxygen in their blood when they are stressed, she simply shuts off her brain…”

这位漂亮的女孩是人类……

this fine lady is a human ...

……在水下待更长时间。

... stay underwater for longer.

“而且不会有压力，所以她可以在水下待更长时间。我明白了。那到底有多久？”

“我不确定，妈妈。但我们初次见面的每一分钟似乎都是美妙的永恒。”

“你肯定对这个女孩着迷了！但你要记住，爱情往往是盲目的……”

“… And doesn't stress, so she can stay underwater for longer. I see. How much longer?”

“I am not sure, Mom, but every minute of our first encounter seemed a wonderful eternity in itself.”

“You surely seem fascinated with this young lady! Just keep in mind that love is often blind …”

“为什么大家都爱这么说？我把眼睛睁得大大的，能看到她每个优雅的动作，她的俯冲和转身，和我摆动鱼鳍是多么合拍。”

“你似乎真的被迷住了，儿子。她没有关闭她的大脑，而是关闭了你的大脑。”

“迷住？这个词到底是什么意思？我只知道，当你们彼此信任时，就不会有焦虑，不会有恐惧。你只要跳舞，享受每一刻。”

“Why do people say that? I had my eyes wide open and could see every one of her elegant movements, her dives and turns, so in harmony with the way I was moving my fins.”

“It seems that you are infatuated, my boy. She did not shut off her brain, she shut off yours.”

“Infatuated? What does that word even mean? All I know is that when you trust each other, there is no anxiety, no fear. You just dance and enjoy every moment.”

她每个优雅的动作……

every one of her elegant movements ...

……曾与鲸鲨游泳……

… swum with whale sharks …

“这就是她能屏住呼吸的时间比任何人都长的原因吗？”

“是的。现在请允许我回味一下与这位女孩共舞的美好瞬间。她曾与鲸鲨游泳，与大白鲨碰过鼻子，还曾与短吻鳄手牵手。”

“这是纯粹的幻想。竟然和鳄鱼手牵手，你不仅盲目地恋爱，你现在还在幻想。这个女孩真的存在吗？”

“And that’s why she is able to hold her breath longer than any other human being?”

“Yip. Now please, allow me to savour my beautiful memories, of dancing with a lady who has swum with whale sharks, has touched her nose to that of a great white shark, and who has even held hands with an alligator.”

“That’s pure fantasy. Holding hands with an alligator, I ask you! You are not only blindly in love you are now also imagining things. Does your lady even really exist?”

“是的，她像我们一样游泳。看不见氧气罐。我告诉你，她是我们中的一员。她很漂亮，她认为我也很漂亮。”

“好吧，如果她是我们中的一员，那么明天我们用气泡钓鱼的时候，就邀请她一起来。”

……这仅仅是开始！……

“Yes, and she swims like us. No oxygen tank in sight. She is one of us, I tell you. And she is beautiful, and she thinks I am too.”

“Well, if she is one of us, let’s invite her to come along when we go fishing with air bubbles tomorrow.”

... AND IT HAS ONLY JUST BEGUN!...

……这仅仅是开始！……

… AND IT HAS ONLY JUST BEGUN! …

Did You Know?

Holding your breath relies on controlling the diaphragm. When we breathe in we contract the muscle of the diaphragm and holding our breath means keeping the diaphragm in a contracted state. Like any muscle, it can be trained to hold on longer.

屏住呼吸依赖于控制隔膜。吸气时，我们收缩隔膜的肌肉，屏住呼吸意味着保持隔膜处于收缩状态。就像任何肌肉一样，可以通过训练来延长屏气的时间。

We can intentionally hold our breath, but anxiety and high stress can also lead to the body adopting shallow breathing, or even altogether stopping to breathe. When faced with danger, the body could freeze, and having consumed all the oxygen, we then faint.

我们可以有意屏住呼吸，但焦虑和高压力也会导致身体浅呼吸，甚至完全停止呼吸。当人面对危险时，身体会紧绷，消耗完体内所有的氧后就会晕倒。

The brain is made up of over 100 billion nerve cells, with each brain cell connected to 10,000 other cells, which equal 1,000 trillion connections in our brain. Every time we learn something, feel an emotion, or practice art, more cells connect.

大脑由1000亿个神经细胞组成，每个细胞又与上万个其他细胞相连，构成了超过1000万亿个神经连接。每当我们学习一些东西，感受一种情绪，或者艺术实践时，更多的细胞连接起来。

The brain consumes power equal to 25 watts of electricity while we are awake. That is enough to power a room with LEDs. When asleep, the power drops to 10 watts. The brain consists of 80% water, which promotes conductivity of electric impulses.

当我们清醒时，大脑消耗的能量相当于25瓦的电力。这足够点亮LED灯为房间照明了。睡眠时，脑能量消耗水平下降到10瓦。大脑的80%是水，水促进了电脉冲的传导。

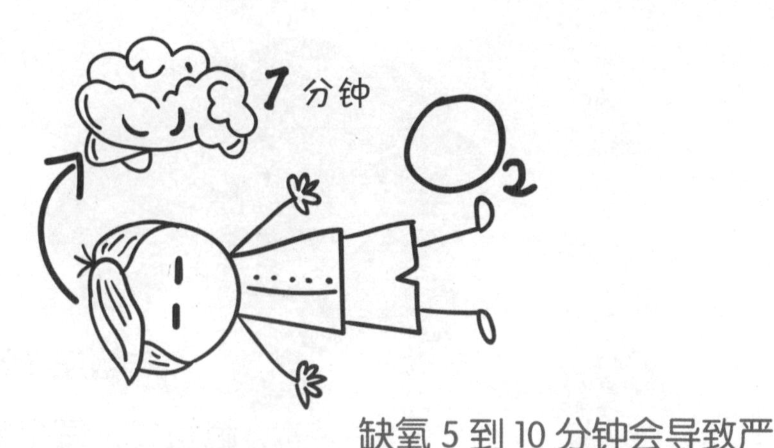

Loss of oxygen for a period of 5 to 10 minutes can cause serious brain damage. The brain can stay alive for only 4 to 6 minutes without oxygen, before cells start dying. We lose consciousness 10 seconds after the blood supply to the brain has stopped.

缺氧 5 到 10 分钟会导致严重的脑损伤。大脑在没有氧气的情况下只能存活 4 到 6 分钟，然后细胞开始死亡。大脑的血液供应停止 10 秒钟后，我们就会失去知觉。

The brain, which in an adult weighs 1.5 kg, consumes at least 20 percent of the body's oxygen. When we are scared or anxious, it could consume as much as 40 percent, causing our natural defence system to shut down, and we lose consciousness.

成人的大脑重 1.5 千克，消耗身体至少 20% 的氧。当我们害怕或焦虑时，脑耗氧量占比可达 40%，我们的固有免疫系统可能因此关闭，甚至会出现意识丧失。

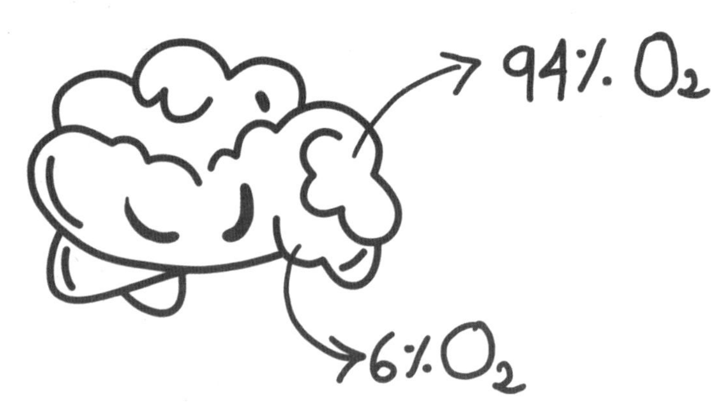

The brain's grey matter consumes up to 94 percent of cerebral oxygen, while the white matter, which makes up fully 60 percent of the brain's mass, consumes only 6 percent. Progressive reductions in oxygen decreases the level of mental alertness.

大脑灰质消耗高达 94% 的脑氧，而占大脑质量 60% 的白质只消耗 6%。氧的不断减少会降低警觉水平。

When in love, the brain releases the same cocktail of neurotransmitters and hormones that are released by amphetamines, leading to increased heart rate, loss of appetite and sleep, and intense feelings of excitement.

恋爱时，大脑会分泌与在安非他明刺激作用下产生的神经递质和激素相同的混合物，导致心率加快、食欲和睡眠下降，以及强烈的兴奋感。

Does it sound crazy to swim and dance with a whale?

与鲸游泳和跳舞听起来是不是很疯狂?

Are there dreams that are impossible, and dreams that are possible?

有没有不可能实现的梦想，有没有可能实现的梦想?

Is dancing with someone a good way to get to know that person?

与某人跳舞是了解他的好方法吗?

Does love make you blind?

恋爱会使你变得盲目吗?

How can you learn to hold your breath for longer? Is it even possible to train your muscles to keep your lungs full of air for longer periods? The secret lies in relaxing, and calming yourself down even if you find holding your breath for long periods very stressful. Please, do not experiment with this in the water. Sit down and list the steps you would have to take to achieve holding your breath for extended periods. Ask a few friends to do the same. Now follow your proposed plan to see if it works. Compare notes with your friends to see what you can learn from each other about extending the length of time you are able to hold your breath. It takes discipline and dedication to control your breathing, but you will learn a skill that may one day come in very useful.

怎样才能学会长时间屏住呼吸呢？有没有可能训练肌肉，让肺部长时间充满空气？秘诀在于放松，长时间屏气很有压力，要让自己平静下来。不要在水里做实验。坐下来，列出需要采取的步骤。请朋友也这样做。现在按照你提出的计划来看看是否有效。和朋友们比较一下，看看能否取长补短。控制呼吸需要付出和自律，不过一旦你掌握了这项技能，说不定将来非常有用。

学科知识

Academic Knowledge

生物学	大脑是身体的控制中心：接收、解释定向的感官信息；大脑更像一个层次丰富的丛林生态系统，而不是一台计算机；脑组织的耗氧量是肌肉组织的三倍；几乎每一种病毒、真菌、寄生虫和细菌（如乳酸菌、弯曲杆菌、梭状芽孢杆菌和拟杆菌）都是厌氧菌，不能在含氧环境中成长，而那些生活在陆地上的细菌则不能在盐水中生存。
化　学	褪黑素帮助调节睡眠模式；大脑需要不断的氧气和葡萄糖供应；氧气是一种天然解毒剂；在细胞水平上由于缺氧而形成癌症；氧分子（O_2）是保持氧气化学性质的最小微粒。
物　理	舞蹈时重心的重要性；为了完成一个完美的转身，重心需要在脚上。
工程学	用无装备潜水技术代替潜水装备。
经济学	新兴的在线约会和相亲服务业以每年25%的速度增长；2018年，网上婚恋有1 500个应用程序或网站，成为一个年产值30亿美元的行业。
伦理学	网上相亲更多的是为了即时满足，而不是作为建立长期关系的过程，其中一半人是为了找乐子。
历　史	巴比伦人最早描述月光对人类的影响；15世纪意大利文艺复兴时期舞蹈在宫廷正式兴起；北美殖民地开拓时期，许多人在寻找配偶之地举办民歌或对舞活动。
地　理	29.5天的月运周期及其对潮汐的影响；美人鱼出现在欧洲、近东、亚洲和非洲的民间传说中。
数　学	舞蹈中的对称和不对称；数学和舞蹈技巧都运用了几何图形，包括正方形、圆形、三角形、菱形和曲线；舞蹈编导运用数学法则来编排轻盈流畅的舞蹈。
生活方式	迷恋是爱情的影子，它使人的生活脱离了现实。
社会学	社区内的人互相分享记忆，这加强了人之间的联系；派对承载并接纳了社会中许多人的脆弱；婚介在相亲中所起的作用。
心理学	焦虑是一种身体、情绪或精神上的紧张状态，影响呼吸模式；焦虑会导致胸部肌肉严重紧张，以至于呼吸中断；派对可以让人尽情享受感官体验；派对作为逃避现实的一种方式。
系统论	将物种间的对话作为一个新前沿，取决于我们是否愿意放弃人类是世界中心的执念；必须建立一个与自然和谐相处，更加可持续的社会。

情感智慧

Emotional Intelligence

鲸妈妈

鲸妈妈理解儿子的兴奋。看到他快乐，她也很高兴，并回想起自己的经历。她渴望知道儿子下一步的行动，很吃惊儿子没有安排与神秘女郎再次约会。她追问了很多问题，想搞清楚女孩的身份。当得知女孩是人类时，鲸妈妈觉得难以理解，认为儿子是一个梦想家。她意识到儿子被迷住了，到了迷恋的程度，担心他心生幻想。起初，她怀疑这位女孩是否存在，但当看到儿子如此投入，并且认为这位女孩是“鲸的一员”时，她接受了这种不寻常的友谊，并要求儿子邀请这位女孩一起去钓鱼。

鲸儿子

鲸儿子乐意与妈妈分享。他让妈妈放心，说这只是一次舞蹈，尽管舞伴非常特别。当他意识到妈妈的担忧时，他缓解了她的焦虑，表示希望能再次见到这个女孩。鲸妈妈根本不信女孩是人类，他就一点点地告诉她细节。妈妈认为这样的邂逅是不可能的，他就平静地告诉她信任在减轻压力和焦虑中的作用。他确定女孩存在，还说她和他们一样游泳，设法让母亲相信。这让鲸妈妈发出了让女孩第二天和他们一起钓鱼的邀请。

艺术

The Arts

舞蹈是空间中的几何运动。从芭蕾到探戈，我们会发现每个动作都被准确量化。先在纸上画几个图形，比如圆或三角形。再画几组线，平行、垂直或相交。你能不能灵活地表现这些形状？对着镜子来做，看看胳膊和腿能不能完全对称地移动。可以考虑用平面或立体的纸或画布来展示舞蹈的构图，以及流畅和优雅的动作。你能用画笔捕捉舞动的感觉，用泥塑来展现舞者的优雅吗？

思维拓展

Systems: Making the Connections

我们经常认为与非人类交流和互动只能在实验室里，由人控制各种条件。当我们进入其他生物的自然环境，例如潜入海洋和鲸一起游泳，像它们一样行动，那么就可以扩展我们的边界，帮助我们了解潜在的科学原理。人在水下时会变得焦虑，大脑耗氧量会达到峰值。这严重限制了我们在水下与其他物种互动的时间。如果我们能从这些焦虑中解脱出来，与水下物种的长时间互动就成为可能。物种间的交流确实是可能的，就像日本潜水冠军二木爱所展示的那样，她曾与鲸互动，也曾与鲨鱼和短吻鳄邂逅。人类会害怕自己不熟悉的动物，尤其是在感觉不安全的环境中遇见又大又危险的动物。人类害怕捕食者，但地球上最大的捕食者难道不是人类自己吗？正是人类的行为导致生物多样性的大规模丧失。人类行为导致气候变化是一个事实。我们需要寻找新的解决方案，而这些方案往往会在我们意想不到的地方有所发现。孩子们有一种天生的与所有生物交流的渴望。对他们来说，与大自然中所有生物打交道是很正常的。这种交往能力将使得我们的社会发展变得更加可持续。

动手能力

Capacity to Implement

掌握呼吸技巧可以拯救生命，无论是你的还是他人的。在水下或烟雾弥漫的房间，如果不能呼吸怎么办？这个练习将帮助你学习更多的屏气技巧，但不要在水里做，否则容易发生危险。先找到隔膜的确切位置，观察吸气和呼气时胸部的变化。你能不能感觉到隔膜的扩张和收缩？现在正常呼吸，然后屏气，用秒表看看能坚持多久。接下来用力吸气，慢慢呼气，多试几次，让血液中充满氧气，然后再次屏气并计时。你能坚持一分钟吗？尽可能放松，毫无压力地重复练习。这是不是对你有些帮助？当你掌握了这门技巧，请把它教给别人。说不定什么时候会用到。

故事灵感来自

This Fable Is Inspired by

二木爱

Ai Futaki

二木爱1980年出生于以和服闻名的日本金泽，3岁时便学会了游泳，从此就一直和水打交道，玩各种水上运动。22岁时，二木爱决定去探索世界，前往洪都拉斯参加了水肺潜水训练。后来，她搬到墨西哥，提升了自己的潜水和摄影技能。在泰国旅行时，她接触到不带氧气罐的深潜。她觉得自己就像美人鱼，并养成了一种能够在水下生活并与水下生物亲密接触的感觉。2011年，《吉尼斯世界纪录大全》收录了她的第一个潜水纪录，不久，她又创造了第二个潜水纪录。她现已开始为探索频道和日本广播公司（NHK）工作。她还担任了保护海狮、海豚和鲸的形象大使，鼓励人们重新与水建立联系。她也是“海洋保护基金会漫游计划”大使，该活动利用一艘通过太阳能电池将海水转化为氢气的帆船进行环球航行，提高世界各国的人们对海洋中塑料的关注。

图书在版编目(CIP)数据

冈特生态童书.第七辑：全36册：汉英对照 /
(比)冈特・鲍利著；(哥伦)凯瑟琳娜・巴赫绘；
何家振等译.—上海：上海远东出版社，2020
ISBN 978-7-5476-1671-0

Ⅰ.①冈… Ⅱ.①冈… ②凯… ③何… Ⅲ.①生态
环境-环境保护-儿童读物—汉英 Ⅳ.①X171.1-49

中国版本图书馆CIP数据核字(2020)第236911号

策　　划 张　蓉
责任编辑 程云琦
封面设计 魏　来李　廉

冈特生态童书
屏住呼吸
[比]冈特・鲍利　著
[哥伦]凯瑟琳娜・巴赫　绘
章里西　译